More Penguin Place Value
Hundreds, Tens, and Ones

Kathleen L. Stone

Enjoy these other books by Kathleen L. Stone

Penguin Place Value
A Math Adventure

Number Line Fun
Solving Number Mysteries

Riley the Robot
An Input/Output Machine

Mason the Magician
Hundreds Chart Addition

Katelyn's Fair Share Picnic
More Math Fun

Money Tree Mysteries
Adventures with Quarters

Alien Even and Alien Odd
A Math Space Adventure

Kenley's Line Plot Graph
Another Math Adventure

Matthew's Sunshine Bakery
Multiplication Arrays

Firefighter Gary's Fire Safety Rules

Samantha's Search
3D Shapes

Grandma's Quilts
Fun with Fractions

From My Quilted Heart to Yours (Book 1)
Heart Warming Quilts and Heart Healthy Recipes for Your Loved Ones

From My Quilted Heart to Yours (Book 2)
Quilts and Blocks from the Children's Book, Grandma's Quilts

ISBN-13: 978-1522998662
ISBN-10: 1522998667

Dedication

To learners everywhere, young and old alike: keep learning, keep questioning, keep dreaming, and keep growing. Education gives you wings … where you fly is up to you!

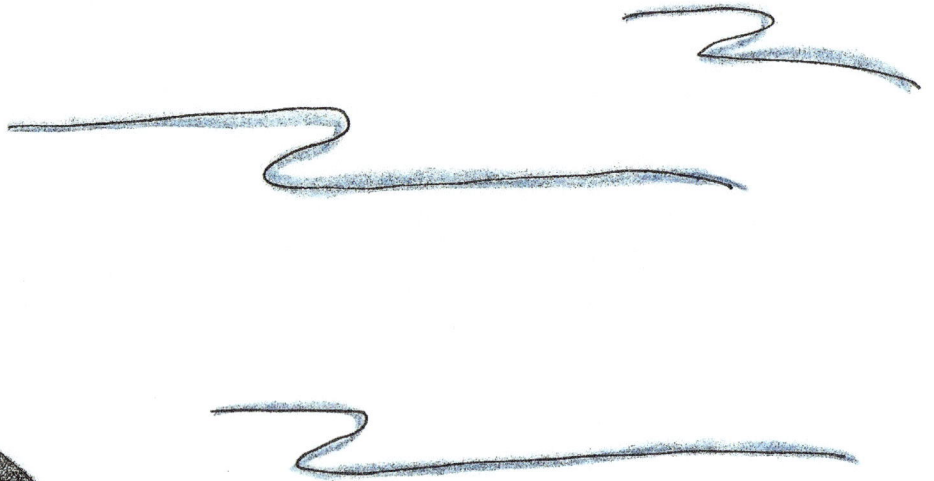

Remember that nice penguin family
Who caught fish all day by the sea?
Well, their business has really been growing.
Come along with me and you'll see.

They place all the fish they catch
On a tray to take back to their store.
But the tray can only hold *nine* fish.
It simply can't hold any more.

Boxes are used to hold *ten* fish
And are kept in a refrigerated van.
Then driven carefully back to their store.
It's really a very good plan.

10   20   30   40   50

60   70   80   90   100

If each van holds *ten* boxes,
Then how many fish can they haul?
If you count by tens you'll find,
There are *one hundred* fish in all.

Let's play a fun little game
And look at all the fish they've caught.
We'll try to figure out exactly
How many fish that they've got!

 = 100

 = 70

 = 8

100 + 70 + 8 = 178

Take a close look at these fish
And tell me how many are in this freight.
One van, *seven* boxes, and *eight* on the tray.
Did you count *one hundred seventy-eight?*

  = 400

     = 60

= 3

400 + 60 + 3 = 463

Four vans, *six* boxes, and *three* on a tray
Are the next group of fish that I see.
How many fish does that make?
You are right! *Four hundred sixty-three.*

 = 200

 = 9

200 + 9 = 209

I told you this would be fun.
You are really doing fine!
Let's look at these fish next.
Did you count *two hundred nine*?

= 600

    **= 20**

   **= 1**

600 + 20 + 1 = 621

Our penguin friends aren't finished.
They've only just begun.
Six vans, *two* boxes, and *one* on the tray.
That makes *six hundred twenty-one.*

= 300

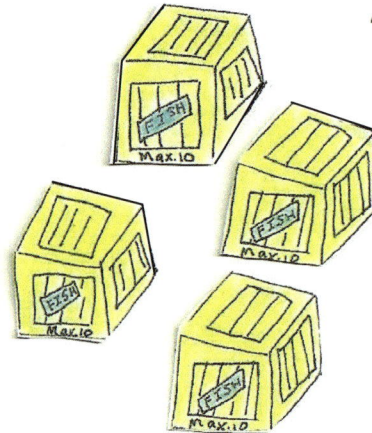

= 40

= 5

300 + 40 + 5 = 345

Look closely at this next group
Of fish going on a drive.
Count carefully and you'll see
That there are *three hundred forty-five*.

 = 500

 = 30

 = 4

500 + 30 + 4 = 534

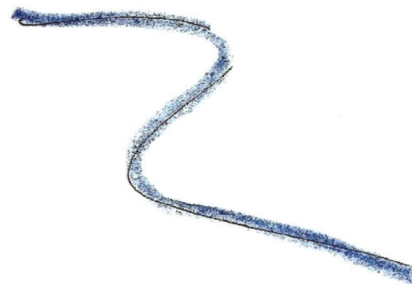

Today's fishing is almost over.
Here's the last fish to be shipped to their store.
Five vans, *three* boxes, and *four* on the tray.
Did you count *five hundred thirty-four*?

The penguins will do more fishing
Once all these fish have been bought.
They would love for you to come back
And count all the fish they have caught!

Place Value and Expanded Notation

Children need to understand and apply concepts of whole numbers. **More Penguin Place Value** provides practice with the concept of **place value** and **expanded notation**. Place value helps us understand the value of a numeral (i.e. **2** has a different value in each of these numbers: **2, 20, 200, 2,000**, etc.). Place value tells us if the **2** stands for two, twenty, two hundred, or two thousand. Place value will also play an important role as children eventually learn how to do addition and subtraction with double- and triple-digit numbers (with and without *regrouping*). Children learn to write numbers in expanded notation form to show the values of the digits in various numbers.

Children should be given many opportunities to work with concrete examples (hands on manipulatives) before moving on to more abstract concepts.

Enrichment Activities

Place Value Jars

Materials needed:

small clear containers
large variety of small objects
place value mat
small paper cups
paper bowls
paper and pencil

Place Value Mat

Clear plastic containers work well with this activity, but you could also use small Ziploc bags. *Oriental Trading* is a great resource for small objects … they have small erasers and other small, thematic toys. Pennies, pasta, metal nuts, etc. can also be used.

Preparing Place Value Jars:

Place a different number of similar objects in each container (i.e. *one hundred three* pennies in one container, *one hundred fifteen* bear erasers in a second, and so on). Be sure and number each container.

Playing the Game:

Children estimate how many objects are in each container and then use their *place value mat* to count them (*10* objects are put into the cup and moved to the "tens" place – *100* objects are placed in a bowl and put into the "hundreds" place). After they have counted the objects, have them write the number in expanded notation form and as a number word.

Place Value Egg Hunt

Materials needed:

plastic eggs
strips of paper
permanent marker
paper and pencil

379

300+70+9=379   three hundred seventy-nine

521

500+20+1=521   five hundred twenty-one

Preparing Place Value Eggs

Using the permanent marker, write numbers on the outside of each plastic egg. Write the expanded notation form and the number word on a slip of paper and place it inside of the correct egg.

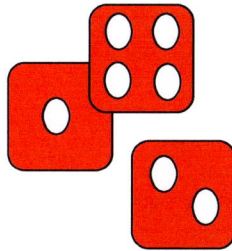

Playing the Game:

Children search for the hidden eggs. Once found, they read the number on their egg and write both the number word and the expanded notation form on their paper. They can self-correct their work by opening their egg and checking their answers.

Mystery Number

Materials needed:

3 dice

Playing the Game:

Before the game begins determine if the player with the largest or smallest number will be the winner. Players take turns rolling the three dice and making a numeral from the three numbers rolled. For added challenge have the players write their number words and the expanded notation form.

How to best use this book … to engage children even more, cover the numbers on the illustrations with *Post It* notes when you read the book aloud to them.

ABOUT THE AUTHOR

Kathleen Stone is a National Board Certified educator and is currently teaching second grade. *More Penguin Place Value* is her fifteenth book published. She loves spending time with her family. She and her husband Gary live in the Olympia area but enjoy traveling across the United States. When not teaching, Kathleen can often be found quilting, sitting by the lake reading, or exploring new parks with her grandchildren!

Math is all around us
No matter where you turn
Open your mind to the wonders of math
And all that you can learn

Printed in Great Britain
by Amazon